Our Star Goes West

The Chaneys bring in Star and her daughter, Starburst, from their pasture in Maryland to load on a truck to go to their new home in America's Heartland. Pictured from left, are, Star, Jaxson, Lee, Rianna, Becky and Sheridan Chaney, and Star's daughter, Starburst.

By Twins Rianna and Sheridan Chaney

Nine-year-old sisters, Rianna and Sheridan Chaney, were a bit nervous when their parents decided they were relocating from a farm in Maryland to a big ranch in Nebraska. The transition was made easier when the girls found out their beloved Hereford cow, Star, Star's daughters, Starburst and Starbucks, and their new calf, Butterscotch, would make the 1,400-mile trip to their new home. "Our Star Goes West" is a sequel to the Star Series — Book #1, "Little Star...Raising Our First Calf" and Book #3, "Star Becomes a Mother." Star's new story traces her move across the country and her new life on a ranch in south central Nebraska.

To all the Farmers and Ranchers in America,

We salute all the hard working farmers and ranchers who spend 365 days a year, seven days a week and 24 hours a day producing safe, wholesome and nutritious food for consumers in the United States and overseas.

Published by Down Under Publications

All rights reserved. No part of this book either in part or in whole, may by reproduced, transmitted, or utilized in any form, by any means, electronic, photographic, or mechanical, including photocopying, recording, or by any information storage system, without permission from the author, except for brief quotations embodied in literary articles and reviews.

First Edition, November, 2013
©Copyright, Rebecca Chaney, 2013
ISBN 978-0-9818468-4-2

Chaneyswalkabout@aol.com
website: www.rebeccalongchaney.com
Blog: chaneytwinsagbooks.blogspot.com
Edited by Rebecca Long Chaney
Photographs by Kelly Hahn Johnson and Rebecca Long Chaney
Layout and Design by Kathy Moser Stowers
Lesson Plans for Books #4 and #5 by Laura Keilholtz, professional educator.

Our favorite Hereford beef cow, Star, on the right, gets a friendly nudge from her daughter, Starburst. We are so excited Daddy is taking Star and Starburst to our new home in America's Heartland.

R.F. PETTIGREW ELEMENTARY

 While Rianna pushes the door of the trailer shut, I peek in at Star to let her know how much we love her. She was an orphaned calf and we raised her on a bottle six years ago. She is such a big girl now, and we take good care of her.

Dad drives out of our farm lane with Star and her family loaded in the livestock trailer to begin the long westward journey. We will join Dad in a few months, but know he will look after our cattle until we arrive at our new home.

 Boy, it sure is cold at our new home. Brrr... even this snowstorm can't stop Dad and me from delivering fresh water to the cattle being fed on cornstalks. The field is filled with the leaves and some corn left over after the farmers harvest their corn crop in the fall. Cornstalks are a good source of nutrients.

What a wonderful surprise when Star gives birth to a heifer calf. She is adorable and we name her Stargage. Star has raised four heifer calves and we know she is going to be a good mother, again.

Our state has been in a two-year drought. In addition to the young cattle grazing on pasture, we supplement them with grain. Dad drives the Cake Truck with distillers grain to feed the animals. Cattle are recyclers because they eat leftover products from food processors and ethanol plants that would otherwise end up in landfills or wasted.

We love to feed the cattle with Dad. We get to flip a switch that Dad calls "the breakfast bell." It's a loud siren, like an ambulance, that lets the cattle know there is grain in the truck and it's time to eat.

Our Dad helps take care of a 300-cow commercial Red Angus herd. For two months in the spring he helps to check cattle day and night for newborn calves. He makes sure the calf is born healthy and puts in an ear tag to identify the calf.

We enjoy doing chores on the ranch that are safe for my sister and me to do. Dad teaches us how to give a nasal vaccination to our Hereford calves to keep them healthy and strong.

R.F. PETTIGREW ELEMENTARY

We love to take our cattle to shows. We learn a lot about responsibility, sportsmanship and teamwork. Showing our two Hereford heifers at the Junior National Hereford Expo in Kansas City, Missouri, is exciting. It's the first time for Star's daughter to be at a national show.

We really enjoy 4-H livestock judging. We judge or evaluate beef cattle, lambs, swine and goats. There are four animals in each class, we take notes on each animal, and place them first through fourth. Judging helps us learn more about livestock and decision making, improves our public speaking skills and builds our self confidence.

 Did you know cattle are wildfire fighters and environmentalists? Grazing reduces the risk of wildfires because cows eat the grass that might be fuel for fires. As cattle walk through pastures their hooves aerate the soil and allow air and water to enter the ground. They also press seeds into the soil which helps establish new growth of grasses and plants.

Ranchers and farmers provide 75 percent of the nation's habitat for wildlife. These mule deer in our neighbor's cornfield are just a few of the animals we enjoy seeing on the ranch. We also see whitetail deer, turkeys, pheasants, badgers, rabbits and lots of birds.

I think she likes me. New life at the ranch is always a miracle. Daddy's first ranch horse, Ladyblue, had a filly. The foal was born black but as she grows she will turn grey like her mother. We hope one day we can get a ranch horse and help gather the cattle.

Branding day is a long work day. First, the cowboys and cowgirls go out early on horseback and gather the 300 cows and calves. It's a big social event, too, with neighbors and friends coming to help.

After the cattle are gathered in the holding pens, the calves and their mothers are separated. The cows are vaccinated and wait in the pasture to reunite with their calves. We help our friends, Johanna and Marie, move the calves to another pen. The sorting sticks we use help guide the calves safely.

We love to watch the cowboys and cowgirls rope calves. They are so awesome. We practice throwing the lasso with our Dad and when we are older, we will help lasso calves, too.

The branding crew work together to make sure all calves are handled with care. Mr. Scott, the ranch owner, gives the calves their brand, or permanent symbol. Brands identify which rancher owns the cattle.

Montella is a real cowgirl. She rides, ropes and helps me hold my first calf. It's a small calf, but I still use all my strength. The cowboys and cowgirls are teaching us so much about ranch life.

Life on the ranch is not all work and no play. On hot days, Mom and Dad take us to the huge cattle watering tanks. The cattle don't even seem to mind that we are cooling off on a sunny afternoon.

We have lots of fun when Johanna and Marie come over for a play day. It seems we always find mud. We are all ranch girls, love cattle and horses, and especially love playing in mud. We walk in it, sit in it, and even slide in it.

Raising a bucket calf is fun, but a lot of work, too. The orphaned calf depends on us to feed him milk and grain for a few months. Then, he will graze or eat grass in a pasture. When he is about 700 pounds he goes to a feedyard to finish growing. Ranchers and farmers take good care of their animals because they provide our bodies with the number one source of protein — beef.

Star and her family sure look good in the pasture at our new ranch home. Star, far left, stands with her two daughters, Stargage and Starburst, and her granddaughter, Stardust. Star is a great cow, has raised four good calves, and has given us special memories to last a lifetime.

Glossary

4-H – The largest youth organization in North America focusing on leadership, citizenship and life skills.
Aerate – Air circulating through the soil.
America's Heartland – A word to describe states located in the middle of the United States.
Branding/brands – Brands are permanent symbols ranchers give their livestock to identify their ranch home.
Cake Truck – Is a utility pickup truck equipped with a grain tank used to feed cattle out on pasture.
Commercial Red Angus – Red Angus is a breed of cattle. Commercial means they are not registered.
Distillers Grain – Is a byproduct or what is left over from making ethanol.
Drought – An extended period of dry weather damaging crops and reducing grass availability to livestock.
Environmentalists – Someone who protects the air, water, land, animals, plants and other natural resources.
Ethanol Plants – A facility that produces ethanol, a type of fuel, from starch-based crops like corn.
Farmers – A farmer is a person who grows crops, like corn, wheat, barley and oats, on a farm.
Feedyard – Once beef cattle reach about 700 pounds, they finish growing to market weight at a feedyard.
Filly – A young female horse.
Foal – A horse that is one year old or younger.
Grazing – Cattle and other livestock graze or eat growing grass.
Habitat – The natural environment or place where wildlife or other organisms live.
Heifer – A female calf that has not yet delivered her first calf.
Hereford – A popular breed of beef cattle that originated in England.
Nasal Vaccination – A vaccine that is given in the nose or nasal cavity.
Orphaned – A child or young animal that has lost its parents and depends on others to care for it.
Ranch/rancher – A rancher is a person who raises livestock on a ranch or large portion of land.
Supplement – Adding feed to the animal's diet to achieve a balanced diet assuring complete health.

Fun Livestock Facts

More than 97% of U.S. beef cattle farms and ranches are family owned • Beef is an excellent source of Zinc, Iron and Protein • Beef fat, proteins and gelatin are ingredients in cosmetics, soap, medical products, marshmallows, toothpaste and more • 800,000 ranchers and farmers raise one million cattle in the U.S. • The top six beef producing states are Texas, Missouri, Oklahoma, Nebraska, South Dakota and Montana • There are more cattle in Nebraska than people • The National Football League requires 3,000 cowhides to make enough footballs for one season • Epinephrine comes from cattle adrenal glands and is used to treat asthma and allergies • Thrombin comes from cattle blood and is used to help clot human blood • The liver extract from cattle is used in treating anemia • Insulin is taken from the pancreas of cattle and is used for the treatment of diabetes • Agriculture is the largest segment of the U.S. economy, and the beef industry is the largest segment of the agricultural economy.

For Teachers and Parents

 www.Factsaboutbeef.com – Funded by beef checkoff dollars. This website debunks myths about beef through science, research and actual facts regarding all aspects of the beef community from the field to the table.

 www.ancw.org – The mission of the American National CattleWomen is to promote and support the industry, and encourage women involved in beef and related agribusiness. They work to sustain the integrity of the beef industry through consumer awareness, education, and promotion; continuing to respect the past, living in the present and looking toward the future.

 SchoolWellness.org – This website provides a variety of tools to help teachers, school wellness leaders, health professionals and families address the nutrition and health challenges facing the youth of today.

 www.beefnutrition.org – Beef nutrition is brought to you by America's Beef Producers through the Cattlemen's Beef Board and State Beef Councils. Events and seminars, media reports, materials and downloads, nutrition information, nutrition research, Dietary Guidelines Alliance, Nutrition Rich Foods Coalition, and recipes are available on this website.

 **www.animalagalliance.org** – It's never too early to start teaching your children about agriculture! The Alliance has compiled a collection of books that provide easy-to-read information about food production and feature fun illustrations of farm and ranch life.

 www.myamericanfarm.org – Built for educators, learners and their families, My American Farm offers free agriculturally themed educational games, downloadable educator resources and fun activities for families to explore.

 BeefItsWhatsforDinner.com – This website is the number one consumer site for beef recipes and information on preparing and serving beef.

 **www.Teachfree.com** – Teach free provides preschool through 12th grade educators with high-quality educational materials that supplement the curriculum. The National Cattlemen's Beef Association works collaboratively with its partners at the state level. There are state beef councils in 45 states, and they may offer free materials and assistance to teachers in their states. Most of the educational materials on Teachfree have been developed as a result of this partnership.

 **www.4-h.org** – 4-H has grown into a community of 6 million young people across America learning leadership, citizenship and life skills. 4-H can be found in every county in every state, as well as the District of Columbia; all U.S. territories and more than 98 countries around the world. 4-H'ers participate in fun, hands-on learning activities, supported by the latest research of land-grant universities.

 **www.FFA.org** – FFA is a dynamic youth organization that is a part of agricultural education programs at middle and high schools with nearly 1/2 million enrolled. Today, student members are engaged in a wide range of curriculum and FFA activities, leading to more than 300 career opportunities in agriculture.

 **www.agclassroom.org** - The Agriculture in the Classroom National Resource Directory is an online searchable database that lists hundreds of educational resources designed to help educators locate high quality classroom materials and information to increase agricultural literacy among their Pre-K through 12th grade students.

 **www.kellyhahnphotography.com** - Kelly Hahn Johnson is an award-winning photojournalist with more than 20 years of experience. See more photos, videos and testimonials here.

 **www.rebeccalongchaney.com** – View photos, testimonials and video footage of Rebecca, Rianna and Sheridan's mission to teach young people about agriculture. Lesson plans for all Chaney books are free and downloadable here and at www.pabeef.org under educational resources.

In 2013, Lee, Becky, Rianna and Sheridan Chaney moved from a farm in Maryland to a ranch in Nebraska. Becky and her daughters started making school appearances engaging elementary students with their agricultural facts and farming story when the girls were only four years old. They are proud their books are among favorites in homes and schools around the country.

The Chaneys are passionate about agriculture and passionate about spreading their message through agricultural education presentations and their books. The girls have settled into their new school which is kindergarten through 12th grade, in one building. They ride a van to school, enjoy fourth grade, play in the youth volleyball league and are learning to play the guitar.

They are members of 4-H and their main projects are livestock, baking and shooting sports. They enjoy 4-H livestock judging and have won several contests in Maryland and Nebraska. Rianna and Sheridan love their new ranch horse, Pryor, and hope to show him in 4-H. On weekends the girls love baking treats for Cowboy Church and spending the afternoon in the pasture checking cows and calves with their parents. For more information about their ag presentations, or to order books, contact Becky at chaneyswalkabout@aol.com, info@rebeccalongchaney.com or check out her website at www.rebeccalongchaney.com

The Chaneys are proud their first four children's books have received numerous "Agricultural Book of the Year" recognitions! The "Chaney Twins' Ag Series" and Lesson plans have been added to the USDA Agriculture in the Classroom Resource Directory as well as the American Farm Bureau's Ag Literacy List.

Lesson Plans are now available free in downloadable form on the Pennsylvania Beef Council website at www.pabeef.org and at www.rebeccalongchaney.com
Box book discounts are available!

Kathy Moser Stowers, greenhouse plants and produce grower, lives with her husband, Tom and son, Wade on her family's second farm, Twin View Acres 2, near Jefferson, Maryland.

She has 25 years of experience in design and layout. She has shared her talents and expertise with local, state, national and international organizations and events.

For aid in designing brochures, pamphlets, program books, business cards, etc., contact Kathy at wakstowers@aol.com or call (301) 748-9112.

Award-winning photographer Kelly Hahn Johnson not only is known for her unique photojournalism style but also her approach to portraiture. She's won countless awards and her images have been featured in local, state and national publications. She loves observing people, moments and emotions, creating photos that will be treasured for years to come, like the priceless images she's captured of the Chaney twins over the years. She lives with her husband Blane and son Brady in a century-old renovated house in Sharpsburg, Maryland.

Visit her online gallery at kellyhahnphotography.com.
Contact Kelly at info@kellyhahnphotography.com or call her at (240) 285-3677.

A special thank you to Scott, Kim, Johanna and Marie Ford of Cross Diamond Cattle Company, and Andy and Megan Nation, for making our first year on the ranch a great adventure!!